International Mathematics

Workbook 3

Andrew Sherratt

HODDER
EDUCATION
AN HACHETTE UK COMPANY

Hachette UK's policy is to use papers that are natural, renewable and recyclable products and mode from wood grown in well-managed forests and other controlled sources. The logging and manufacturing processes are expected ta conform to the environmental regulations of the country of origin.

Orders: please contact Hachette UK Distribution, Hely Hutchinson Centre, Milton Road, Didcot, Oxfordshire, OX11 7HH. Telephone: +44 (0)1235 827827. Email education@hachette.co.uk. Lines are open from 9 a.m. to 5 p.m., Monday to Friday. You can also order through our website: www.hoddereducation.com

© Andrew Sherratt 2009
First published in 2009 by
Hodder Education, an Hachette UK Company,
Carmelite House, 50 Victoria Embankment,
London EC4Y 0DZ

Impression number   11
Year   2022

Cover photo © PhotoAlto Agency RF/Punchstock
Illustrations by Macmillan Publishing Solutions
Typeset in 12.5/15.5pt Garamond by Macmillan Publishing Solutions
Printed and bound by CPI Group (UK) Ltd, Croydon, CR0 4YY

A catalogue record for this title is available from the British Library

ISBN 978 0 340 96750 8

Contents

Graphs of linear equations

1 Plot these four graphs on the **same** Cartesian plane.

    a) $y = 2x$ **b)** $y = 2x + 1$

    c) $y = 2x - 1$ **d)** $y = 2x + 2$

2 Plot these four graphs on the **same** Cartesian plane.

    a) $y = -2x$ **b)** $y = -2x + 1$

    c) $y = -2x - 1$ **d)** $y = -2x + 2$

3 Choose a suitable scale and plot five points to draw the graph of the equation $y = 5x - 4$.

Use the graph to find:

    a) the value of y when $x = -1.5$, $x = 0.6$, $x = 2.2$

    b) the value of x when $y = -6.5$, $y = 0.5$, $y = 3.5$.

4 Choose a suitable scale and plot five points to draw the graph of the equation $y = 3x + 5$.

Use the graph to find:

    a) the value of y when $x = -2$, $x = 0.6$, $x = 1.5$

    b) the value of x when $y = -1$, $y = 0.8$, $y = 2.9$.

5 Calculate the gradient of the line segment joining each of these pairs of points:

    a) $A(4, 3)$ and $B(1, 2)$ **b)** $C(2, 3)$ and $D(-2, -1)$

    c) $E(4, -1)$ and $F(2, 5)$ **d)** $G(3, -1)$ and $H(-2, 4)$

    e) $J(-1, 2)$ and $K(-5, 4)$ **f)** $L(0, -3)$ and $M(-3, 0)$

6 Calculate the gradient of each of these lines.

    a)

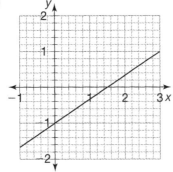

b)

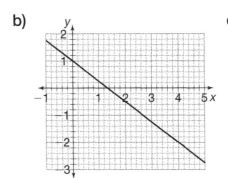

c)

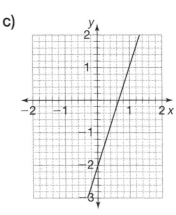

7 The points in each set lie on the same straight line. Write down the equation of each line.

a) $(-5, -5), (-3, -5), (0, -5), (4, -5), (7, -5)$

b) $(-1, 8), (-5, 8), (3, 8), (10, 8), (12, 8)$

c) $\left(-7, \frac{1}{3}\right), \left(-\frac{3}{4}, \frac{1}{3}\right), \left(0, \frac{1}{3}\right), \left(4, \frac{1}{3}\right), \left(8, \frac{1}{3}\right)$

d) $(-2, 14), (-2, 9), (-2, 1), (-2, -2), (-2, -5)$

e) $(1.1, 3), (1.1, 2), (1.1, 1), (1.1, 0), (1.1, -1)$

f) $\left(-\frac{1}{2}, 2\right), \left(-\frac{1}{2}, 3\right), \left(-\frac{1}{2}, 4\right), \left(-\frac{1}{2}, 5\right), \left(-\frac{1}{2}, 6\right)$

g) $(1, 0.2), (2, 0.2), (3, 0.2), (4, 0.2), (5, 0.2)$

h) $(0, 10), (0, -10), (0, 11), (0, -11), (0, 12)$

i) $\left(-\frac{5}{8}, \frac{1}{2}\right), \left(-\frac{5}{8}, \frac{1}{4}\right), \left(-\frac{5}{8}, \frac{2}{3}\right), \left(-\frac{5}{8}, \frac{1}{3}\right), \left(-\frac{5}{8}, 1\right)$

j) $(2, 11), (4, 11), (6, 11), (8, 11), (10, 11)$

8 Write down the coordinates of three points on each of these lines.

a) $y = 7.5$ b) $x = 8\frac{1}{2}$

c) $x = -10\frac{2}{3}$ d) $y = 21$

9 Draw the graphs of these equations on the same Cartesian plane.

a) $x = 3.6$ b) $y = -2\frac{1}{2}$

c) $y = 4.5$ d) $x = -\frac{6}{5}$

10 For each of these straight lines, write down
i) the gradient and ii) the y-intercept:

a) $2y + x = 4$ b) $y - x = 0$

c) $x + y = 4$ d) $4x + 3y = 9$

e) $3y + 6 = x$ f) $4 - 3y = 2x$

g) $3x + 2y = 6$ h) $-x + 4y = 20$

i) $4y - 3x = 8$ j) $4x - y = 6$

k) $-7x - 4y = 16$ l) $9x + 3y = 1$

m) $4x + 3y = 8$ n) $5x - 9y = -7$

o) $12x = 2y + 1$ p) $y - 2 = 0$

11 i) Write down the equation of the straight line that has the gradient and y-intercept given.

ii) Write each equation using only positive whole number coefficients and constants.

	Gradient	y-intercept			Gradient	y-intercept
a)	-5	-5		**b)**	$1\frac{1}{2}$	$\frac{1}{4}$
c)	$\frac{5}{8}$	$-\frac{3}{4}$		**d)**	$2\frac{3}{8}$	0
e)	0	$3\frac{1}{3}$		**f)**	$-\frac{2}{3}$	$-1\frac{1}{4}$

12 Work out the equation of each of these lines.

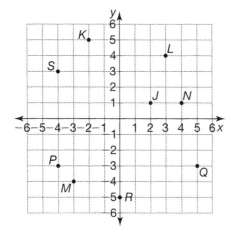

a) JK **b)** LM

c) NS **d)** PQ

e) RQ

13 Write down the equation of the line that is parallel to the given line and cuts the y-axis at the given point.

	Equation	Point			Equation	Point
a)	$y = -\frac{1}{2}x + 3$	$(0, 5)$		**b)**	$y = x + 3$	$(0, 2)$
c)	$2x - 3y = -\frac{1}{2}$	$(0, -2\frac{1}{2})$		**d)**	$3y = -x + 1$	$(0, 3)$
e)	$y - 4x = 5$	$(0, 2.6)$		**f)**	$x + y = 3$	$(0, -1)$
g)	$x - 5y = 3$	$(0, 8)$		**h)**	$-2x - 3y = 6$	$(0, -\frac{3}{4})$

14 Work out the equation for each straight line, given the gradient and one point on the line.

	Gradient	Point			Gradient	Point
a)	-3	$(1, 4)$		b)	3	$(-4, -7)$
c)	$\frac{1}{2}$	$(6, 1)$		d)	$1\frac{1}{3}$	$(-2, 0)$
e)	4	$\left(-1, \frac{1}{2}\right)$		f)	$\frac{1}{3}$	$(-3, 1)$
g)	0	$\left(-5, \frac{3}{4}\right)$		h)	undefined	$\left(-\frac{5}{8}, \frac{2}{3}\right)$

15 Write down the equation of the line that is parallel to the given line, and passes through the given point.

	Equation	Point		Equation	Point
a)	$y = -2x + 3$	$(1, 2)$	b) $2x + 3y + 4 = 0$	$(-3, 1)$
c)	$y - 3 = 0$	$(3, -2)$	d) $y = x + 1$	$(0, 2)$
e)	$3y = x + 6$	$(-1, 5)$	f) $4y + x = 2$	$(-2, 4)$

16 Find the equation of the line that passes through each of these pairs of points.

a) $(0, -4)$ and $(4, 0)$ b) $(0, 1)$ and $(3, -8)$

c) $(6, 5)$ and $(4, 3)$ d) $(2, 5)$ and $(4, 2)$

e) $\left(\frac{1}{2}, 1\right)$ and $\left(\frac{1}{4}, \frac{1}{2}\right)$ f) $(3, -1)$ and $(-6, -4)$

g) $(-1, 2)$ and $(3, 4)$ h) $(3, -1)$ and $(-3, 5)$

Linear inequalities

1 Copy the statements and write '>', '=' or '<' in each space to make the statement true.

a) 15 _____ 1.5

b) -5 _____ -7

c) 0 _____ -0.2

d) -0.3 _____ -0.4

e) $\sqrt{5}$ _____ 5

f) 5.1^2 _____ $(-5.1)^2$

g) $(-6)^2$ _____ $(-5)^2$

h) $(-6)^3$ _____ $(-5)^3$

i) 9^2 _____ -9^2

j) $\frac{1}{3}$ _____ $\frac{1}{2}$

k) $-\frac{1}{3}$ _____ $-\frac{1}{2}$

l) $\left(\frac{1}{3}\right)^2$ _____ $\left(\frac{1}{2}\right)^2$

m) $\left(-\frac{1}{3}\right)^2$ _____ $\left(-\frac{1}{2}\right)^2$

n) $\left(-\frac{1}{3}\right)^3$ _____ $\left(-\frac{1}{2}\right)^3$

o) $(0.3)^2$ _____ 0.3

p) $(0.4)^2$ _____ 0.16

q) 0.3 _____ $\sqrt{0.3}$

r) $7 + y$ _____ $5 + y$

s) $10 - x$ _____ $8 - x$

t) $2y + 2$ _____ $2y - 3$

2 Copy each statement and write '>', '=' or '<' in each space to make the statement true.

a) If $x > -30$ and $-30 > y$, then x _____ y

b) If $x > y$ and $y > 8.2$, then x _____ 8.2

c) If $y + 7 = x$ then x _____ y

d) If $y = 9 - x$ and $x > 0$, then 9 _____ y

e) If $x > y$, then $14y$ _____ $14x$

f) If $x > y$, then $-\frac{x}{45}$ _____ $-\frac{y}{45}$

g) If $y > x$, then $\frac{5}{8}x$ _____ $\frac{5}{8}y$

h) If $x > y$, then $\left(-\frac{2}{3}\right)y$ _____ $\left(-\frac{2}{3}\right)x$

i) If $x > 0$ and $y < 0$, then $-xy$ _____ 0

j) If $y < 0$, then $\frac{y}{-17}$ _____ 0

k) If $y - 11 = x$, then x _____ y

l) If $x < y$ and $x > 0$, then y _____ 0

3 Solve each of these inequalities and show the solution on a number line.

a) $2x + 36 < 4$

b) $6 + 11x > -60$

c) $-9x + 5 \geqslant -58$

d) $42 > 3x + 3$

e) $4x - 68 > -4$

f) $37 \leqslant 17 - 2x$

g) $-3 - 7x > -17$

h) $14 < 5x + 34$

i) $6x - 4 < -40$

j) $30 - 8x < 6$

k) $-28 \geqslant 12x - 4$

l) $-\frac{x}{2} + 20 \leqslant 4$

m) $-18 > \frac{x}{6} - 10$

n) $-\frac{3}{2}x + 9 \leqslant 24$

o) $\frac{3}{10}x + 21 < 0$

p) $\frac{2}{3}x + 4 < 16$

4 Solve each of these inequalities and show the solution on a number line.

a) $5x + 2 > 3x + 10$

b) $19 - 3x \leqslant x - 4$

c) $4x + 49 < 9 - x$

d) $8 - 6x \geqslant 4x - 22$

e) $3(x - 4) > 15$

f) $15 - 2x \leqslant 9 + 3x$

g) $3(2x + 1) \geqslant 4x + 9$

h) $5 - 4x \leqslant -3 - x$

i) $2(7x - 1) \geqslant 3(5 - x)$

j) $4 - 3x < 5 - 4x$

k) $5(1 - 2x) \geqslant 6(2 - x)$

l) $7(2x - 4) < 3(5 - 3x)$

m) $4(1 - 3x) - 14 > 4(2x + 3) - 9x$

n) $5(2x + 1) \geqslant 6(3x - 5) - 2(x + 7)$

5 Solve each of these inequalities.

a) $\dfrac{2 - 3x}{3} < -\dfrac{5}{6}$

b) $\frac{1}{3}(x + 9) \geqslant -\frac{1}{2}(2x - 5)$

c) $\dfrac{x + 2}{5} \geqslant \dfrac{3 - x}{4}$

d) $\dfrac{x - 1}{4} - \dfrac{x - 1}{3} < \dfrac{7}{12}$

e) $\dfrac{x + 1}{2} + \dfrac{x - 1}{3} \leqslant 1$

f) $\dfrac{2x + 1}{3} < \dfrac{3x - 4}{5} + \dfrac{2}{3}$

g) $\dfrac{11 - x}{4} > 2 - x$

h) $\dfrac{2x - 3}{6} + \dfrac{x + 2}{3} < \dfrac{5}{2}$

i) $\dfrac{2x + 7}{2} \geqslant \dfrac{5x - 3}{4} + \dfrac{7x - 1}{8}$

j) $0.3(x - 1) > 0.7(2x - 3)$

6 Find the solutions for each pair of inequalities and show the solutions on a number line.

If there are no solutions, say so.

a) $2x + 5 > 1$ and $3x + 4 > 7$

b) $3x - 5 < 27$ and $25 \leqslant 4x - 3$

c) $x - 2 \leqslant 6$ and $4x - 3 > 18$

d) $-2x + 8 < 14$ and $3x + 1 < 1$

e) $x + 1 < 14$ and $2x + 5 > 14$

f) $x - 4 \leqslant 3$ and $3x \geqslant -6$

g) $5(x + 4) \geqslant 5$ and $2(x + 4) < 12$

h) $4 + 3x > 0$ and $16 - 3x > 1 - x$

i) $2x - 5 \geqslant 1$ and $3x - 1 > 26$

j) $3x < 2x - 3$ and $7x > 4x - 9$

7 Write down the integer values of x that are true for each of these double inequalities.

a) $5 < 4x + 1 \leqslant 13$

b) $-4 < 3x + 2 \leqslant 11$

c) $-5 < 3(x + 5) < 0$

d) $6 \geqslant 4x > 3x - 1$

e) $x + 1 < 27 < x + 4$

f) $7 \leqslant 2x + 1 \leqslant 21$

g) $4x + 5 \leqslant 5x - 2 < 4x + 7$

h) $x + 1 < 11 < 2x + 6$

i) $20 - 2x > 2x + 4 > 1 - x$

j) $\frac{1}{2}x + \frac{1}{5} \geqslant \frac{2}{5}x > x - 5$

k) $4x + 5 \leqslant 5x - 2 \leqslant 4x + 7$

l) $\frac{1}{2}x + 6 < \frac{1}{4}x + 10 < x + 5$

1 Solve these quadratic equations by factorising (if necessary, rearrange the equation).

a) $(x - 3)(x + 5) = 0$

b) $(x - 4)(2x + 3) = 0$

c) $(t + 5)(2t - 7) = 0$

d) $6y + y^2 = 0$

e) $m^2 - 7m = 0$

f) $3x^2 - 3 = 0$

g) $2d^2 - 50 = 0$

h) $64 - b^2 = 0$

i) $n^2 - \frac{1}{4} = 0$

j) $a^2 + 6a - 27 = 0$

k) $p^2 + 7p = 60$

l) $1 + 3y = 10y^2$

m) $4 - 3b - b^2 = 0$

n) $12 - 7x + x^2 = 0$

o) $t^2 - 15t - 54 = 0$

p) $b^2 + 15b + 54 = 0$

q) $x^2 - 12x + 32 = 0$

r) $y^2 - 14y - 176 = 0$

s) $z^2 - 26z + 133 = 0$

t) $c^2 + 2c - 195 = 0$

u) $x^2 + 5x - 24 = 0$

2 Solve these equations by factorising the quadratic expressions.

a) $2a^2 + 13a + 6 = 0$

b) $4b^2 - 12b - 91 = 0$

c) $6c^2 - 37c - 35 = 0$

d) $8d^2 + 18d - 35 = 0$

e) $10g^2 - 123g - 91 = 0$

f) $12h^2 + 119h + 99 = 0$

g) $18k^2 - 95k + 25 = 0$

h) $33m^2 + 76m - 32 = 0$

i) $5n^2 + 68n + 39 = 0$

j) $7p^2 + 31p + 12 = 0$

k) $9q^2 - 45q + 56 = 0$

l) $11r^2 + 145r - 126 = 0$

m) $36s^2 + 3s - 5 = 0$

n) $32t^2 - 30t - 27 = 0$

o) $44y^2 - 43y - 15 = 0$

3 Solve these quadratic equations by factorising (if necessary, rearrange the equation).

a) $(x + 3)^2 = 16$

b) $12x^2 - x = 6$

c) $8x^2 - 22x = 63$

d) $x^2 + 4 = 8x - 8$

e) $(x - 2)(x + 2) = 12$

f) $9x^2 - 3x = 20$

g) $(x - 2)^2 = 9(x - 2)$

h) $8x^3 = 15x - 14x^2$

i) $6(x - 1)^2 = 16 - 8x$

j) $6x^2 = x + 15$

k) $x^2 = 10x + 24$

l) $x(2x + 5) = 3$

m) $6x^3 - x^2 = x$

n) $(6x + 5)(x - 1) = -3$

o) $4(x^2 - 2x - 3) = 5(x - 3)$

4 Find an equation (with whole number coefficients) for each solution set.

a) $\{5, -6\}$ b) $\{3, 4\}$

c) $\{-2, -5\}$ d) $\left\{7, -\frac{1}{2}\right\}$

e) $\{0, 12\}$ f) $\left\{-\frac{3}{4}, 1\right\}$

g) $\left\{1\frac{1}{2}, -1\frac{1}{4}\right\}$ h) $\left\{\frac{7}{8}, -\frac{5}{6}\right\}$

i) $\left\{2\frac{1}{2}, 1\frac{3}{4}\right\}$ j) $\left\{-1\frac{1}{3}, \frac{5}{8}\right\}$

5 Fifteen less than the square of a number is the same as twice the number. Find the number.

6 The sum of a number and twice its square is 6. Find the number.

7 Find two consecutive positive odd integers that have a product of 35.

8 Find two consecutive even integers so that the square of the bigger integer, decreased by twice the smaller integer, is 52.

9 Find two consecutive positive odd numbers so that the sum of their squares is 74.

10 The sum of the squares of two consecutive even integers is 340. Find the two numbers.

11 Find two consecutive positive odd integers so that the square of their sum exceeds the sum of their squares by 126.

12 Two positive numbers differ by 7 and the sum of their squares is 169. Find the numbers.

13 The sum of the squares of three consecutive positive integers is 245. Find the biggest number.

14 Find three consecutive positive integers so that the square of the first, increased by the last, is 22.

15 The length of a rectangle is 4 cm more than the width. The area is 96 cm².

Find the length and width of the rectangle.

16 The sides of a square are increased by 3 m. The area of the new square is 64 m².

Find the length of a side of the original square.

17 A rectangular garden measures 4 m by 5 m. The length and the width are both increased by the same amount. The area of the garden is now 56 m².

How big is the new garden?

18 A rectangular field measures 70 m by 50 m. There is a path all the way around the outside of the field. The area of this path is 1024 m².

How wide is the path?

19 The area of a rectangular field is 450 m². The length is 7 m longer than the width.

Find the perimeter of the field.

20 The length of a rectangle exceeds its width by 8 cm. The length is now halved and the width is increased by 6 cm. The area of the rectangle is now 36 cm² less than it was.

Find the perimeter of the original rectangle.

21 The width of a rectangle is $(2x + 1)$ cm and the length is $(3x + 1)$ cm. The area of the rectangle is 117 cm².

Find the perimeter of the rectangle.

22 The hypotenuse of a right-angled triangle measures $5x$ cm. The other two sides measure $(5x - 1)$ cm and $(x + 2)$ cm.

Calculate the area and the perimeter of this triangle for each value of x.

23 Sarah is 3 years younger than her sister, Meg. The product of their ages is 154.

How old is Sarah?

24 Add a constant term to each expression to make it into a perfect square.

a) $x^2 + 14x$ b) $x^2 - 22x$

c) $x^2 + \frac{2}{3}x$ d) $x^2 - 3x$

e) $49x^2 + 42x$ f) $9x^2 - 30x$

g) $16x^2 - 4x$ h) $25x^2 + 17x$

25 Change each of these quadratic expressions into the completed square form.

a) $x^2 + 20x + 96$ b) $x^2 - 6x - 72$

c) $x^2 + 5x - 14$ d) $x^2 - 15x + 44$

e) $36x^2 + 60x + 50$ f) $64x^2 + 32x - 12$

g) $9x^2 - 3x - 30$ h) $100x^2 - 14x + 1$

26 Solve each of these equations by completing the square.

If the answers are not whole numbers, give them correct to 2 decimal places.

a) $x^2 - 6x = 72$ **b)** $x^2 + 24x + 140 = 0$

c) $x^2 + 195 = 28x$ **d)** $x^2 + 4x + 1 = 0$

e) $4x^2 - 20x = 0$ **f)** $100x^2 + 20x = 0$

g) $x^2 + 10 = 7x$ **h)** $x^2 + 5x = 20$

 i) $x^2 + 7x + 9 = 0$ **j)** $x^2 + 40 = -15x$

k) $8x^2 + 32x + 14 = 0$ **l)** $2x^2 + 7x + 2 = 0$

m) $3x^2 + 8x + 2 = 0$ **n)** $5x^2 + 12x + 3 = 0$

o) $5x^2 + 30x = 18$ **p)** $x^2 = 11x + 26$

q) $x^2 = 7x + 30$ **r)** $2x^2 - x - 3\frac{1}{2} = 0$

s) $x^2 = 2x + 5$ **t)** $5x^2 + 2 = 16x$

 27 Solve these quadratic equations using the quadratic formula.

Give your answers correct to 2 decimal places.

a) $x^2 + 2 = 5x$ **b)** $x^2 + 4x + 2 = 0$

c) $x^2 + 7x + 3 = 0$ **d)** $x^2 + x = 1$

e) $x^2 = 8x + 5$ **f)** $x^2 = 5x + 3$

g) $x^2 + 6x = 1$ **h)** $3x^2 = 4x + 5$

 i) $2x^2 + 4x = 1$ **j)** $2x^2 + 4 = 7x$

k) $3x^2 = 7$ **l)** $3x^2 - 2x = 20$

m) $3x^2 + 7 = 12x$ **n)** $4x^2 = 3x + 8$

o) $10x^2 = 9x + 8$ **p)** $3x + 4 = 2x^2$

 Give the answers to questions **28** to **34** correct to 1 decimal place.

28 The length of a rectangle is 3 times the width. The area of the rectangle is $108\,cm^2$.

Calculate the length and width of the rectangle.

29 A city park is in the shape of a rectangle. The length of the park is 2 km less than twice the width.

The area of the park is $9\,km^2$.
Calculate the length and width of the park.

30 The base of a triangle is 3 cm longer than its perpendicular height. The area of the triangle is 35 cm².

Calculate the perpendicular height of this triangle.

31 A farmer decides to make his square field bigger. He adds 3 m to the length and 2 m to the width.

The new area of the field is 90 m².
Calculate the length of a side of the original field.

32 A rectangular table is covered with 600 square tiles. The same table could also be covered by 400 square tiles that have sides 1 cm longer.

Calculate the length of the side of the smaller tile.

33 The length of a rectangle is 6 cm more than the width. The area of the rectangle is 11 cm².

Calculate the length and width of the rectangle.

34 A square picture has a frame that is 1 cm wide all round. The area of the picture alone is $\frac{2}{3}$ of the area of the picture and the frame together.

Find the length of a side of the picture.

Unit 4 Algebraic fractions

1 Simplify each of these algebraic fractions.

a) $\dfrac{x - 3}{7x - 21}$

b) $\dfrac{49 - x^2}{x - 7}$

c) $\dfrac{x^2 - 25}{3x - 15}$

d) $\dfrac{-x^2 + 8x - 16}{3x^2 - 12x}$

e) $\dfrac{n^2 + 7x + 10}{n^2 + 2n - 15}$

f) $\dfrac{6 - 5c - c^2}{7c^2 - 7}$

g) $\dfrac{n^2 + 7n - 18}{n^2 - 4}$

h) $\dfrac{n - 6}{n^2 - 6n}$

i) $\dfrac{-10c^3 - 5c^2}{2c^2 + 15c + 7}$

j) $\dfrac{b^2 + 4b - 21}{2b^2 - 18}$

k) $\dfrac{-x^2 + 8x - 12}{x - 2}$

l) $\dfrac{b^2 + 4b + 4}{2b^2 + 3b - 2}$

m) $\dfrac{c^2 d + 4cd^2}{-c^2 + 16d^2}$

n) $\dfrac{-3c^2 + 6cd}{-3c^2 + 7cd - 2d^2}$

o) $\dfrac{c^2 + cd - 6d^2}{c^2 - 3cd + 2d^2}$

p) $\dfrac{(x + y)^2 - 4y^2}{x^2 - 9y^2}$

q) $\dfrac{5x^2 - 25x}{3x^3 - 75x}$

r) $\dfrac{x^2 + 5x - 24}{3 - x}$

s) $\dfrac{49x - x^3}{7 - 6x - x^2}$

t) $\dfrac{6a^2 - 30a + 36}{4a - 12}$

u) $\dfrac{a^4 - 8a^3 b}{a^3 - 64ab^2}$

v) $\dfrac{4a^2 + 8ab - 12b^2}{6a^2 - 12ab + 6b^2}$

w) $\dfrac{3ab^3(a - 1)}{6a^4 b^4(1 - a)}$

x) $\dfrac{ab^6(a^2 - 2a - 15)}{a^7 b^5(5 - a)}$

y) $\dfrac{3a^3(16 - a^2)}{12a^6(a^2 - 9a + 20)}$

z) $\dfrac{15a^5 b(5 - a)}{6a^2 b^3(a - 5)}$

2 Simplify these algebraic expressions and give the answers in their lowest terms.

a) $\dfrac{x^3}{2y^2} \times \dfrac{6y^4}{xy}$

b) $\dfrac{4-a}{5a} \times \dfrac{a^2+5a}{a^2+a-20}$

c) $\dfrac{x-1}{4xy^3} \times \dfrac{6x^2y}{1-x}$

d) $\dfrac{8a-40}{40-3a-a^2} \times \dfrac{a-8}{2a^2-8a}$

e) $\dfrac{x^2+7x+12}{x-5} \times \dfrac{2x-10}{x+3}$

f) $\dfrac{6a+24}{2a^2+5a-12} \times \dfrac{4a^2-9}{15a^2}$

g) $\dfrac{12x+48}{6x-15} \times \dfrac{4x^2-25}{x^2+9x+20}$

h) $\dfrac{2a^2-13a+15}{8a^2-12a} \times \dfrac{6a-4a^2}{a^2-10a+25}$

i) $\dfrac{2x^2+5x-7}{x+4} \times \dfrac{x^2+4x}{x^2-2x+1}$

j) $\dfrac{m^2}{m^2-7m} \div \dfrac{1}{m^2-4m-21}$

k) $\dfrac{n^2-9n+20}{6m^7n^2} \div \dfrac{5n-20}{10mn^2}$

l) $\dfrac{12n-36}{9-n^2} \div \dfrac{8n^5}{n^2+3n}$

m) $\dfrac{17mn^3}{m^2+2m-35} \div \dfrac{34m^8n^4}{m^2+7m}$

n) $\dfrac{4n^3-25n}{3n^2-16n+5} \div (10n+25)$

o) $\dfrac{m^2-mp}{p^2-pq} \div \dfrac{p^2-mp}{mp-mq}$

p) $\dfrac{x+4}{2x^2-14x} \times \dfrac{x^3+4x^2}{3x-24} \div \dfrac{x^2+8x+16}{x^2-3x-28}$

q) $\dfrac{(2x-5)^3}{3-x} \div \dfrac{2x^2-3x-5}{6x^2+15x} \times \dfrac{x^2-2x-3}{4x^2-25}$

r) $(75x^2-12) \div \left(\dfrac{35-2x-x^2}{x^2+7x} \div \dfrac{x-5}{5x^3+2x^2} \right)$

s) $\dfrac{x^2+18x+80}{x^2+15x+56} \div \dfrac{x^2+5x-50}{x^2+6x-7} \div \dfrac{x-1}{x-5}$

3 Simplify these algebraic expressions and give the answers in their lowest terms.

a) $\dfrac{4b}{5} + \dfrac{b+3}{2}$

b) $\dfrac{2x-y}{6} + \dfrac{y-x}{10}$

c) $\dfrac{2-5n}{2} - \dfrac{3}{5}$

d) $\dfrac{m+2}{6} - \dfrac{2m-3}{8}$

e) $\dfrac{1}{3x} + \dfrac{5}{4x^3}$

f) $\dfrac{7}{10x^2} + \dfrac{1}{2x^3} + \dfrac{11}{5x}$

g) $\dfrac{2}{3} - \dfrac{1}{y} - \dfrac{4}{y^2}$

h) $\dfrac{5}{y} - 4y$

i) $\dfrac{a+6}{4a^3} + \dfrac{4a+3}{12a^2}$

j) $\dfrac{13}{18b^2} - \dfrac{11}{12b^4}$

k) $\dfrac{9}{42x^2y^2} - \dfrac{6}{49x^3y}$

l) $\dfrac{3a+b}{ab^2} + \dfrac{5a-2b}{a^2b}$

4 Simplify these algebraic expressions and give the answers in their lowest terms:

a) $4 + \dfrac{5a}{2a-3}$

b) $\dfrac{7}{x-3} + \dfrac{4}{x^2-9}$

c) $\dfrac{3}{d-7} - \dfrac{2}{3d+1}$

d) $\dfrac{8}{5d+4} - \dfrac{1}{2d-3}$

e) $\dfrac{x}{x+2} + \dfrac{x}{x-2} - 5$

f) $3 + \dfrac{2x}{x-2} - \dfrac{5x}{x-5}$

g) $\dfrac{a-9}{5a+2} - 6$

h) $\dfrac{x-20}{x^2-4x} + \dfrac{x}{x-4}$

i) $\dfrac{7x}{x^2-9x+14} - \dfrac{4}{x-7}$

j) $\dfrac{3}{x-4} - \dfrac{x-9}{x^2-16}$

k) $\dfrac{11m}{m^2+3m-28} + \dfrac{m}{m+7}$

l) $\dfrac{6}{a^2-4} + \dfrac{2}{a+2} + \dfrac{5}{a-2}$

m) $\dfrac{5}{x+5} - \dfrac{2x+5}{x^2+9x+20}$

n) $\dfrac{8k-1}{5k^2-30k+45} + \dfrac{3k-4}{k^2+4k-21}$

o) $\dfrac{5y-30}{y^2-12y+36} + \dfrac{7}{9y} - \dfrac{15y}{3y^2-18y}$

5 Solve each of these equations for the unknown variable.

a) $\dfrac{7}{2x} = 3$

b) $\dfrac{4}{x} + \dfrac{2}{3x} = 1$

c) $\dfrac{5}{x - 4} - 3 = 0$

d) $\dfrac{6}{z + 3} = \dfrac{5}{z - 4}$

e) $\dfrac{4}{b - 3} = b - 3$

f) $\dfrac{2}{5x} = \dfrac{4}{x - 1}$

g) $\dfrac{y + 2}{y} = \dfrac{y}{y + 1}$

h) $\dfrac{5}{a + 4} - \dfrac{2}{a - 2} = 0$

i) $\dfrac{3}{1 + 2x} = \dfrac{5}{3 + 4x}$

j) $\dfrac{x + 2}{6} + \dfrac{x}{3} = \dfrac{2 - x}{2}$

k) $\dfrac{5}{6x} + \dfrac{6}{7x} - \dfrac{9}{14x} = 4$

l) $\dfrac{3}{x + 1} = \dfrac{8}{x + 2} - \dfrac{5}{x + 3}$

m) $\dfrac{3}{2 - x} + \dfrac{5}{4 - 2x} - \dfrac{1}{x - 2} = 4$

n) $\dfrac{3t}{(t - 1)(3t - 2)} - \dfrac{5}{(t - 1)} = \dfrac{3}{3t - 2}$

o) $\dfrac{n}{n^2 - 2n - 8} + \dfrac{1}{n^2 + n - 2} = 0$

p) $\dfrac{4}{a^2 - a - 2} + \dfrac{3}{a^2 - 4} = \dfrac{2}{a^2 + 3a + 2}$

1 Solve each of these pairs of simultaneous equations by equating expressions.

a) $y = 2x + 1$
$y = x - 1$

b) $2x + y = 12$
$x + y = 7$

c) $5x - 2y = 12$
$3x + 2y = 12$

d) $2x = 3y + 4$
$2x = 5y + 8$

e) $5x - 2y = 13$
$3x + 2y = 3$

f) $3x - 2y = 9$
$2x - 2y = 7$

g) $3x - 4y = 2$
$3x = 7y - 1$

h) $-x + 2y = 13$
$x + y = 8$

i) $4x + 5y = 1$
$4x - y = -5$

j) $3x + 2y + 7 = 0$
$5x - 2y + 1 = 0$

k) $3y = 2x - 3$
$5y = 2x - 1$

l) $6x + 5y = 2$
$2x - 5y = -26$

2 Solve each of these pairs of simultaneous equations by the elimination method.

a) $x + y = 16$
$x - y = 0$

b) $3x - y = 1$
$x + y = 3$

c) $x + y = 2$
$x - y = 4$

d) $x - y = 6$
$x + y = 12$

e) $2x + y = 7$
$x + y = 4$

f) $x + y = -1$
$3x - y = 5$

g) $x + 2y = 13$
$x + 4y = 21$

h) $3x - 2y = 5$
$2y - 5x = 9$

i) $5x + 4y = 22$
$5x + y = 13$

j) $3x - 5y = 13$
$5y + x = 7$

k) $2x + 3y = 8$
$2x + y = -4$

l) $7x + 2y = 33$
$3y - 7x = 17$

m) $2x + 3y = -19$
$7x - 3y = 1$

n) $3y - 2x + 15 = 0$
$2x - 2y + 19 = 0$

o) $3x + 2y = 8$
$2y - 5x = 8$

p) $4x + 2y = 24$
$x - 2y = 1$

3 Solve each of these pairs of simultaneous equations by the elimination method.

a) $4x + 3y = 9$
$x + y = 2$

b) $x + 3y = 10$
$2x + 5y = 18$

c) $2x + y = 3$
$3x - 2y = 1$

d) $4x - y = -2$
$3x + 2y = -7$

e) $x + 6y = 0$
$3x - 2y = -10$

f) $x = 3y - 2$
$9y = 4x - 7$

g) $3x - 2y = 11$
$x - 7y = 29$

h) $2x + 3y = 8$
$3x + 2y = 7$

i) $3x = 7 + y$
$5x - 9y = 41$

j) $-3x + 4y = 4$
$9x - 2y = 3$

k) $3x + 4y = 23$
$2x + 5y = 20$

l) $7y + 24x = 9$
$11y - 6x = 36$

m) $3x + 2y = 6$
$5x + 3y = 11$

n) $5x - 4y = 24$
$2x = y + 9$

o) $\frac{1}{2}x + \frac{1}{3}y = 3$
$\frac{1}{4}x + \frac{2}{3}y = 3$

p) $9x = 4y - 20$
$5x = 6y - 13$

4 Solve each of these pairs of simultaneous equations by the substitution method.

a) $x + 5y = 18$
$x = 4y$

b) $x + 2y = 15$
$y = 2x$

c) $y = 3 - 4x$
$6x - 5y = -2$

d) $5x + 6y = 34$
$y = x + 2$

e) $3x + 7y = 4$
$2x + y = -1$

f) $x + 4y = 32$
$4 = 2y - x$

g) $4x + y = 15$
$7x - 3y = 50$

h) $2x - y = 4$
$x + 2y = 7$

i) $5x + 3y = 11$
$4x - y = 2$

j) $4x + y = -12$
$2x + 5y = -6$

k) $3x + y = 11$
$x - y = 1$

l) $6x - 5y = 10$
$5x + y = -2$

m) $3x - 4y = 30$
$2x - 7y = 33$

n) $4x - 3y = 8$
$3x + 5y = 6$

o) $2x - 3y = 13$
$3x - 12y = 42$

p) $5x + 5y = 3$
$10x - 15y = -4$

5 Solve each of these pairs of simultaneous equations.
Choose any method you have learned that you think will be suitable.

a) $x + 2y = 5$
$3x + 4y = 11$

b) $3x - 4y = 10$
$x + 2y = 5$

c) $3x - 2y = 25$
$2x + 5y = 4$

d) $3x + 4y = 5$
$-2x + 5y = 12$

e) $2x + 3y = 19$
$5x + 2y = -2$

f) $3x + 7y = 22$
$5x - 3y = 0$

g) $5x + 3y = 2$
$2y = 13 - x$

h) $3x + 5y = 30$
$2x - 3y = 1$

6 Solve each of these pairs of simultaneous equations by drawing their graphs.

a) $3x + y = 2$
   $2x - y = 3$

b) $y = x - 2$
   $y = -2x + 4$

c) $x + y = 6$
   $y = x - 2$

d) $y = x - 4$
   $x + y = 6$

e) $x + 4y = 12$
   $4x + y = 18$

f) $x + y = 8$
   $y - x = 2$

g) $x + y = 2$
   $2x - y = -5$

h) $3x - 4y = 10$
   $5x + 7y = 3$

i) $x = 3$
   $x - y = 5$

j) $x + 2y = 8$
   $2x + y = 7$

k) $3x - 2y = 13$
   $2x + 2y = 0$

l) $2x - y = 3$
   $3x + 2y = 1$

7 The sum of two numbers is 90 and their difference is 18.

Find the two numbers.

8 Find two numbers so that twice the larger number plus the smaller one is 0, and three times the larger number less the smaller one is 5.

9 The second of two numbers is 4 more than the first. The sum of the numbers is 56.

Find the numbers.

10 The sum of two numbers is 40. If 2 is added to the larger number, the result is equal to twice the smaller number.

What are the two numbers?

11 Find two numbers so that the sum of twice the first and three times the second is 19, and three times the first number less the second one is 1.

12 When both the numerator and denominator of a fraction are decreased by 1, the fraction has a value of $\frac{3}{4}$. When both the numerator and denominator of the same fraction are increased by 1, the value of the fraction is $\frac{4}{5}$.

Find the original fraction.

13 The number of calories in a fruit bun is 20 less than three times the number of calories in a slice of chocolate cake. The bun and the slice of cake together have 500 calories.

How many calories are in each?

14 Three groups of pupils go to a café. The first group pay $5.90 for 7 cups of tea and 6 cups of coffee. The second group pay $9.70 for 18 cups of coffee and 5 cups of tea.

What do the third group pay for 7 cups of coffee and 6 cups of tea?

15 Two CDs and three DVDs cost $31. Three CDs and two DVDs cost $29.

Find the cost of each CD and each DVD.

16 A clothing factory has 540 workers. The cutters are paid $20 a day and the machinists are paid $24 a day. The total wage bill for this factory is $12 000 per day.

How many cutters and machinists work at this factory?

17 Four years ago, Claire was three times as old as Bob. Four years from now, Claire will be only twice as old as Bob.

What are the ages of Claire and Bob now?

18 The sides of an equilateral triangle measure $(5x - 8y)$ cm, $(3y - x + 8)$ cm and $(2x - y + 1)$ cm.

Calculate the perimeter of the triangle.

Unit 6   The geometry of circles

1 Calculate the missing values for each of the sectors in this table. Use $\pi = \frac{22}{7}$.

	Radius of circle	Diameter of circle	Circumference of circle	Area of circle	Angle of sector	Length of arc	Area of sector	Perimeter of sector
a)		28 mm						58.8 mm
b)				3850 cm²	36°			
c)					150°		9240 m²	
d)					280°	44 m		
e)	14 cm							111.6 cm
f)			44 mm			8.8 mm		
g)				3850 cm²			2695 cm²	
h)	27 m				140°			

2 Calculate the missing values for each of the sectors in this table. Use $\pi = 3.142$. Give your answers to one decimal place.

	Radius of circle	Diameter of circle	Circumference of circle	Area of circle	Angle of sector	Length of arc	Area of sector	Perimeter of sector
a)	5 mm				72°			
b)				113.1 cm²			66 cm²	
c)			75.4 cm			61.6 cm		
d)	4 mm							30.5 mm
e)					11.8°	1.6 m		
f)					165.8°		117.2 cm²	
g)				615.8 cm²	79.8°			
h)		40 m						92.4 m

3 Calculate the size of each of the angles marked with a letter.

a)
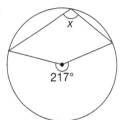

b)
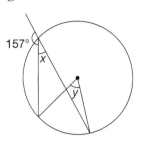

c)
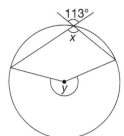

d)
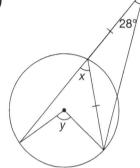

e)
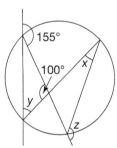

f)
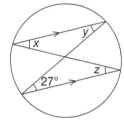

g)
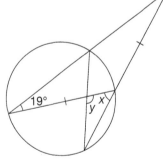

h)
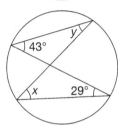

i)
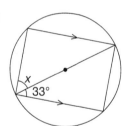

j)
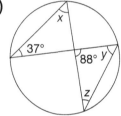

k)
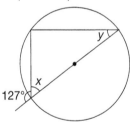

l)
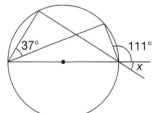

m)
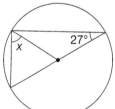

n)
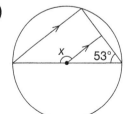

o)
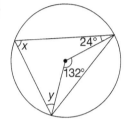

p)
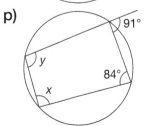

q)
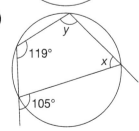

r)
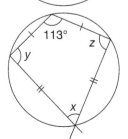

s)

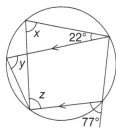

t)

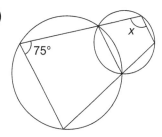

4 Calculate the size of each of the angles marked with a letter.

a)

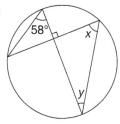

b)

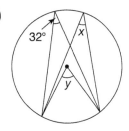

c)

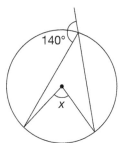

d)

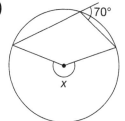

e)

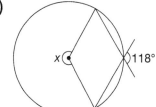

f)

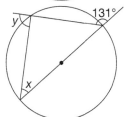

g)

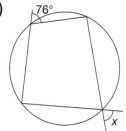

h)

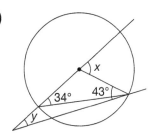

i)

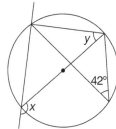

j)

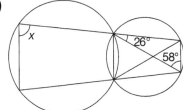

k)

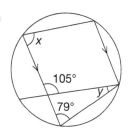

l)

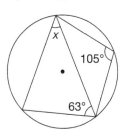

m)

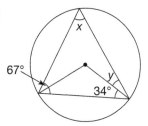

n)

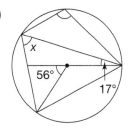

o)

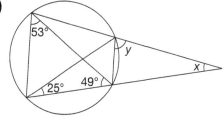

5 Calculate the size of each angle marked with a letter.

a)

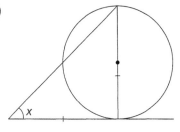

b)

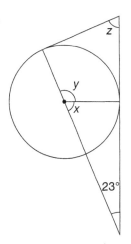

c)

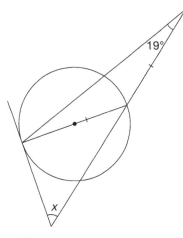

d)

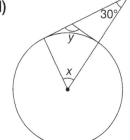

e)

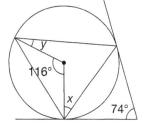

f)

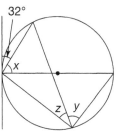

g)

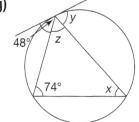

h)

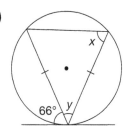

i)

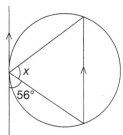

j)

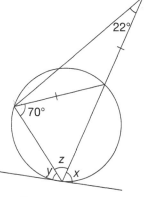

k)

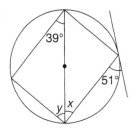

l)

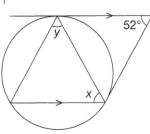

m)

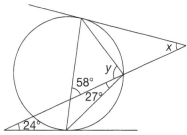

n)

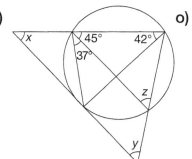

o)

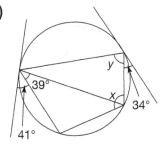

Unit 7 Transformations

1 Write these translations using column vectors in the form $\begin{pmatrix} x \\ y \end{pmatrix}$.

a) 4 right, 5 up **b)** 3 left, 5 up

c) 5.5 right, 3 down **d)** 2 left, 3 down

e) 14 left **f)** 2 right, 1 down

2 Describe in words the translations shown in these column vectors. (Use 'left', 'right', 'up', 'down'.)

a) $\begin{pmatrix} 1 \\ 1 \end{pmatrix}$ **b)** $\begin{pmatrix} 6 \\ 3 \end{pmatrix}$

c) $\begin{pmatrix} -5 \\ 6 \end{pmatrix}$ **d)** $\begin{pmatrix} 1.2 \\ -1.1 \end{pmatrix}$

e) $\begin{pmatrix} -3 \\ -1 \end{pmatrix}$ **f)** $\begin{pmatrix} 0 \\ -7 \end{pmatrix}$

g) $\begin{pmatrix} -2 \\ 0 \end{pmatrix}$ **h)** $\begin{pmatrix} 0 \\ 0 \end{pmatrix}$

3 Use column vectors to describe the translations of these points as shown.

a)

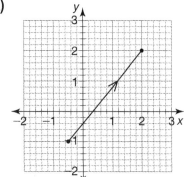

b)

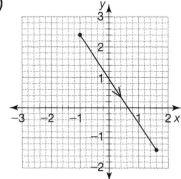

c)

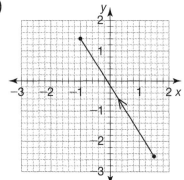

d)

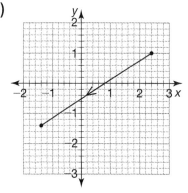

4 Find the coordinates of the image of each of these points after translation along the vector $\begin{pmatrix} -5 \\ -6 \end{pmatrix}$.

a) $(3, -6)$ **b)** $(5, 8)$

c) $(8, 3)$ **d)** $(7, -4)$

5 The point $(4, -2)$ is mapped onto the point $(-1, -5)$ along the translation vector V. Find the coordinates of the image of each of these points after translation along the same vector, V.

a) $(3, -4)$ **b)** $(-5, 0)$

c) $(-3, -3)$ **d)** $(-4, -1)$

6 Copy the figure P onto graph paper (or squared paper).

Draw the image figure after translation along each of these vectors.

a) $V_1 = \begin{pmatrix} 2 \\ -3 \end{pmatrix}$ **b)** $V_2 = \begin{pmatrix} -7 \\ -3 \end{pmatrix}$

c) $V_3 = \begin{pmatrix} -6 \\ 4 \end{pmatrix}$ **d)** $V_4 = \begin{pmatrix} 1 \\ 5 \end{pmatrix}$

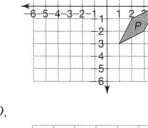

7 The diagram shows quadrilateral $ABCD$. Write down the coordinates of the image quadrilateral after $ABCD$ has been translated along each of these vectors.

a) $\begin{pmatrix} 9 \\ 5 \end{pmatrix}$ **b)** $\begin{pmatrix} -6 \\ 8 \end{pmatrix}$

c) $\begin{pmatrix} 14 \\ -11 \end{pmatrix}$ **d)** $\begin{pmatrix} -7 \\ -15 \end{pmatrix}$

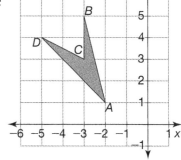

8 a) Write down the vector that will map the shaded kite onto each of the labelled image kites.

b) Write down the vector used to map kite E onto kite C.

c) Write down the vector used to map kite B onto kite E.

d) Write down the vector used to map kite D onto kite A.

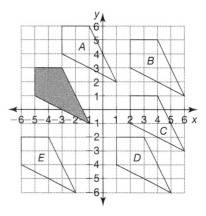

9 The diagram shows the image of a figure after translation along the vector $\begin{pmatrix} -9 \\ -2 \end{pmatrix}$.

Write down the coordinates of the object figure before translation.

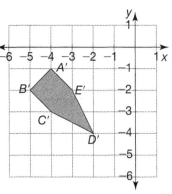

10 a) Find the coordinates of the image of the quadrilateral *PQRS* with *P*(3, 1), *Q*(5, 3), *R*(0, −6) and *S*(−4, −4) after translation along the vector $\begin{pmatrix} 7 \\ -3 \end{pmatrix}$.

b) The image from part **a)** is then translated along $\begin{pmatrix} -2 \\ -5 \end{pmatrix}$. Find the coordinates of this new image.

c) Write down a vector that would map *PQRS* directly onto the image from **b)**.

11 Write down the coordinates of the image of each of these points after reflection in the *x*-axis.

  a) (9, 1)   **b)** (−3, −8)

  c) (0, −7)   **d)** (3, 0)

12 Write down the coordinates of the image of each of these points after reflection in the *y*-axis.

  a) (9, 1)   **b)** (−3, −8)

  c) (0, −7)   **d)** (3, 0)

13 Write down the coordinates of the image of each of these points after reflection in the line $x = -1$.

  a) (5, 6)   **b)** (−6, −1)

  c) (0, −9)   **d)** (4, 0)

14 Write down the coordinates of the image of each of these points after reflection in the line $y = -1$.

  a) (5, 6)   **b)** (−6, −1)

  c) (0, −9)   **d)** (4, 0)

15 Write down the coordinates of the image of each of these points after reflection in the line $y = x$.

  a) (2, 5)   **b)** (−5, −10)

  c) (0, 4)   **d)** (−2, 0)

16 Write down the coordinates of the image of each of these points after reflection in the line $y = -x$.

a) (2, 5) **b)** (−5, −10)

c) (0, 4) **d)** (−2, 0)

17 Write down the coordinates of the image of each of these points after reflection in the y-axis followed by reflection in the line $y = x$.

a) (8, 3) **b)** (−7, −2)

c) (0, 1) **d)** (−6, 0)

18 Write down the coordinates of the image of each of these points after reflection in the x-axis followed by reflection in the line $y = -x$.

a) (8, 3) **b)** (−7, −2)

c) (0, 1) **d)** (−6, 0)

19 Copy each figure onto squared paper and draw its reflection in each of these lines.

  i) the x-axis **ii)** the y-axis

iii) $y = 2$ **iv)** $y = -x$

Write down the coordinates of the vertex B after each reflection.

a)
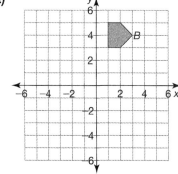

b)
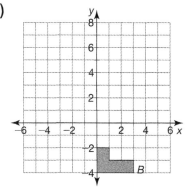

c)
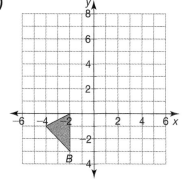

d)
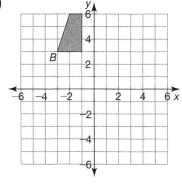

20 Use a construction on squared paper to work out the coordinates of the image for each of these rotations.

a) The point (0, 3) is rotated 90° clockwise about the origin.

b) The point (7, −2) is rotated 90° anticlockwise about the origin.

c) The point (−4, −9) is rotated 180° about the origin.

d) The point (2, 1) is rotated 270° clockwise about the origin.

e) The point (−1, 1) is rotated 180° about the origin.

f) The point (3, 4) is rotated 90° clockwise about the origin.

21 Use a construction on squared paper to work out the coordinates of the image for each of these rotations.

a) The point (5, 1) is rotated 90° clockwise about the point (2, 1).

b) The point (−4, −1) is rotated 90° anticlockwise about the point (−1, 2).

c) The point (3, 4) is rotated 180° about the point (3, 1).

22 Construct the image of each of these line segments after the rotation given and write down the coordinates of A' and B'.

a) 90° clockwise about the origin   **b)** 180° about the origin

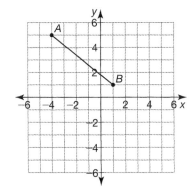 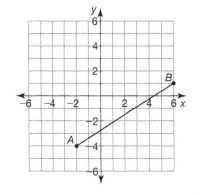

c) 90° anticlockwise about the point (−4, −1)   **d)** 180° about the point (−1, 1)

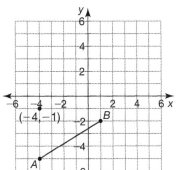 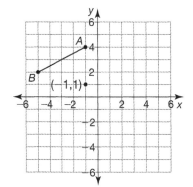

29

23 Construct the image of each of these figures after the rotation given.
Write down the coordinates of the vertex B' after each rotation.

a) 90° clockwise about the origin

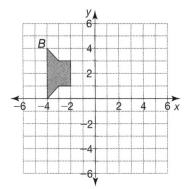

b) 180° about the origin

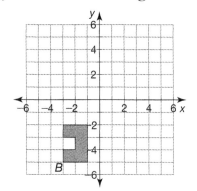

c) 90° anticlockwise about the
point $(1, -3)$

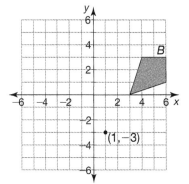

d) 180° about the point
$(-1, -1)$

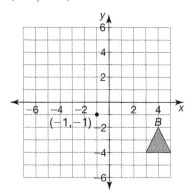

24 Copy each of these figures onto squared paper and then enlarge it
using the centre of enlargement and scale factor given.

a) Centre $(0, 4)$

Scale factor $2\frac{1}{2}$

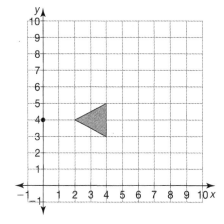

b) Centre $(7, 5)$

Scale factor $1\frac{1}{2}$

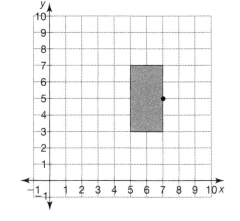

c) Centre (1, 2)

Scale factor $\frac{1}{2}$

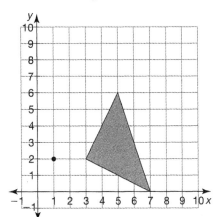

d) Centre (3, 9)

Scale factor $\frac{1}{3}$

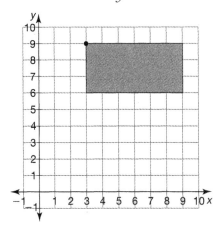

25 Draw on squared paper the rectangle $ABCD$ with vertices $A(0, 0)$, $B(4, 0)$, $C(4, 2)$ and $D(0, 2)$.

Enlarge this rectangle with the origin as the centre of enlargement and a scale factor of $2\frac{1}{2}$.

Write down the coordinates of the image rectangle $A'B'C'D'$.

26 The diagram shows the kite $ABCD$.

Copy this kite onto squared paper.

Work out the coordinates of the vertex C after each of these enlargements.

a) Centre (0, 0), scale factor 2

b) Centre (0, 0), scale factor 3

c) Centre (0, 2), scale factor 2

d) Centre $A(1, 2)$, scale factor 2

e) Centre $B(4, 3)$, scale factor $3\frac{1}{2}$

f) Centre $D(2, 5)$, scale factor $2\frac{1}{2}$

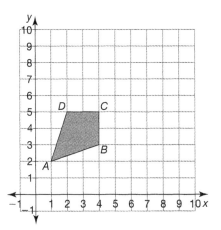

27 For each of these diagrams, find the scale factor and the coordinates of the centre of enlargement.

a)

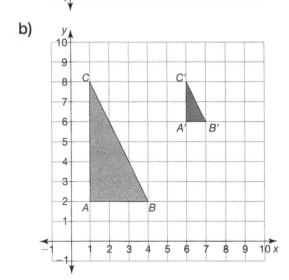

b)

c)

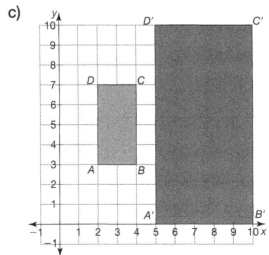

Unit 8

Geometry in three dimensions

1 Copy and complete this table by calculating the missing values for each of the cuboids.

	Length	Width	Height	Surface area	Volume
a)		3 cm	2 cm		24 cm³
b)	7 m	4 m	5 m		
c)	10 cm		5 cm	190 cm²	
d)	45 mm	24 mm			19 440 mm³
e)		5.8 cm	10.6 cm	612 cm²	
f)	3.6 m	3.6 m	4.5 m		
g)	8 cm		4.5 cm		180 cm³

2 Calculate the volume of each of these prisms.

a)

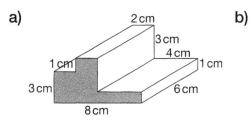

b)

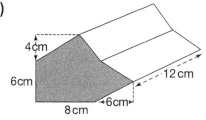

c) Use $\pi = 3.14$.

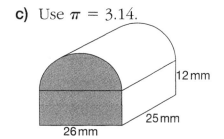

d)

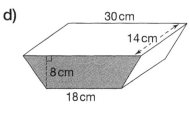

3 Copy and complete this table by calculating the missing values for each of the cylinders correct to 1 decimal place (use $\pi = 3.14$).

	Radius	Diameter	Height	Area of curved surface	Total surface area	Volume
a)	3 cm		7 cm			
b)	4 cm					753.6 cm³
c)		10 m		361.1 m²		
d)			150 mm			105 975 mm³
e)	2.5 cm			314 cm²		
f)		3.5 cm				96.2 cm³
g)		7.2 m	8.5 m			

4 A large water pipe in the shape of a cylinder is made of plastic that is 5 cm thick. The pipe is 3 m long and the inside diameter is 50 cm. Use $\pi = 3.14$ to calculate (correct to 1 d.p.):

   a) the curved surface area on the inside of the pipe

   b) the curved surface area on the outside of the pipe.

 5 For each of these solid figures, draw the net and calculate its total surface area. Use $\pi = 3.14$.

   a)

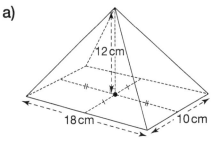

   b)

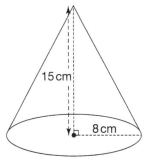

   c)

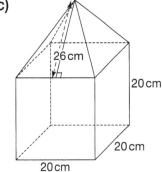

6 Copy and complete the table by calculating the missing values for each of these pyramids.

	Height	Base area	Volume
a)	12 cm	99 cm²	
b)	11 cm		231 cm³
c)		8 cm²	42 cm³

7 Copy and complete the table by calculating the missing values for each of these square pyramids.

	Length of side in base	Height	Base area	Volume
a)	9 cm	8 cm		
b)	5 cm			100 cm³
c)			256 cm²	2304 cm³

8 A pyramid has a right-angled triangle as a base. The two shorter sides of the base measure 5 cm and 12 cm. If the volume of the pyramid is 160 cm³, find its height.

 9 Copy and complete the table by calculating the missing values for each of these cones. Use $\pi = 3.14$ and write your answers correct to 1 d.p.

	Base radius (cm)	Base diameter (cm)	Height (cm)	Slant height (cm)	Base area (cm²)	Volume (cm³)	Curved surface area (cm²)	Total surface area (cm²)
a)	6		8	10				
b)					201	1004.8		628
c)			20	25	706.5			
d)				34		15 072	3202.8	
e)		24				4340.7		1627.8
f)	14.4		27					2034.7
g)			11	8		700.3		
h)			4.2	18			514.3	

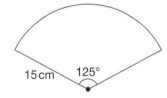 **10** The sector of a circle shown in the diagram is folded to form a cone with an open base.

   a) Calculate the radius of the base of this cone correct to 1 d.p.

   b) Calculate the height of the cone correct to 1 d.p.

   c) Calculate the volume of the cone correct to 1 d.p. (Use $\pi = 3.14$.)

11 The diagram shows two metal curtain poles. They are both the same length of 2.6 m. Pole A is a cylinder of diameter 5 cm and length 2.5 m with a cone on each end.

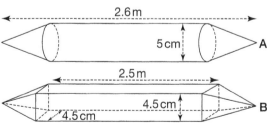 Pole B is also 2.5 m long, has a square cross section with sides 4.5 cm and has a pyramid on each end.

Calculate the volume of each pole to find out which one contains the least amount of metal. (Use $\pi = 3.14$.)

12 Copy and complete the table, using $\pi = 3.14$ to calculate the missing values for each sphere.

Give the radii correct to 1 d.p. and the volumes and surface areas correct to 2 d.p.

	Radius	Diameter	Surface area	Volume
a)		12.0 mm		
b)			2826.00 m²	
c)				4186.67 cm³
d)	1.0 cm			
e)			2461.76 mm²	
f)				179.50 cm³
g)		11.6 cm		
h)			1994.03 m²	

 13 Copy and complete the table, using $\pi = 3.14$ to calculate the missing values for each solid hemisphere.

Give the radii correct to 1 d.p. and the volumes and surface areas correct to 2 d.p.

	Radius	Curved surface area	Total surface area	Volume
a)	7.7 mm			
b)				1630.05 m³
c)		615.50 cm²		
d)	18.4 cm			
e)				6649.52 mm³

14 Fifty-four solid metal hemispheres, each with a diameter of 2 cm, are melted together to make one large sphere.

Calculate the radius of the large sphere. Use $\pi = 3.14$.

15 I have 1 tonne of metal. 1 cm³ of this metal weighs 7.5 g.
How many small metal balls (to the nearest ball) can I make if each has a diameter of 9 cm? Use $\pi = 3.14$.

 16 A hollow cylinder of radius 9 cm contains water to a height of 13 cm. A solid metal sphere is placed in the cylinder and the height of the water goes up by 2 cm.

Calculate the radius of the metal sphere, correct to 2 d.p.
Use $\pi = 3.14$.

17 A water storage tank is in the shape of a
cylinder with a hemisphere on top. It has
a diameter of 4.8 m and the cylinder part
has a height of 16.4 m.

The tank contains only $\frac{1}{4}$ of its total
volume of water.

Work out the height of the water in
the tank. Use $\pi = 3.14$.

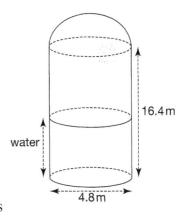

18 A solid hemisphere of diameter 36 cm has
two other hemispheres cut out as shown
in the diagram.

Use $\pi = 3.14$ to calculate:

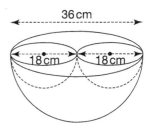

a) the amount of water that this shape holds

b) the mass of the solid shape to the
nearest gram if the hemisphere is made
of metal that weighs 5 g per cm³.

19 A sphere of diameter 30 cm is half-full of oil. This oil is transferred
to a cylinder with a diameter of 18 cm.

What is the depth of the oil in the cylinder, correct to 1 d.p.?
Use $\pi = 3.14$.

20 A solid figure is made from a hemisphere, a cylinder and a cone
joined together as shown in the diagram.

All three shapes have the same radius of 4 cm.

The volume of the cone is $1\frac{1}{2}$ times the volume
of the hemisphere.

The height of the cylinder is $1\frac{1}{2}$ times the height
of the cone.

Use $\pi = 3.14$ to calculate:

a) the height of the cone

b) the height of the cylinder

c) the total volume of the solid, correct to 2 d.p.

1 For each equation, copy and complete the table of x- and y-values, and then plot and draw the graph of the equation. Use the graph to read off the given values.

a) $y = 10 - x - x^2$

x	-4	-3	-2	-1	0	1	2	3
y								

Estimate:

i) the value(s) of y when $x = -2.5$ and 2.2

ii) the value(s) of x when $y = 0$ and 7.5.

b) $y = x^2 - 2$

x	-4	-3	-2	-1	0	1	2	3	4
y									

Estimate:

i) the value(s) of y when $x = -1.5$

ii) the value(s) of x when $y = 0$.

c) $y = x^2 + 3x - 2$

x	-5	-4	-3	-2	-1	0	1	2
y								

Estimate:

i) the value(s) of y when $x = 0.5$ and -2.5

ii) the value(s) of x when $y = 0$ and -1.5.

d) $y = 3 + 13x - 4x^2$

x	-2	-1	0	1	2	3	4	5
y								

Estimate the value(s) of x when $y = 0$ and -10.

2 For each equation, copy and complete the table of x- and y-values, and then plot and draw the graph of the equation.
Use the graph to read off the given values.

a) $y = x^3 - 2x - 1$

x	-3	-2	-1	0	1	2	3
y							

Estimate:

i) the values of y when $x = -2.5$, 0.5 and 2.2

ii) the values of x when $y = 0$, -10 and 15.

b) $y = \frac{1}{2}(x^3 + 6)$

x	-3	-2	-1	0	1	2	3
y							

Estimate the value(s) of x when $y = 4$.

c) $y = 5 + 6x - x^3$

x	-3	-2	-1	0	1	2	3
y							

Estimate:

i) the values of y when $x = -2.5$, 0.5 and 2.2

ii) the values of x when $y = 0$, 8 and -2.

d) $y = 3x - x^3$

x	-3	-2	-1	0	1	2	3
y							

Estimate:

i) the value(s) of y when $x = 1.7$

ii) the value(s) of x when $y = -6.6$.

3 For each equation, copy and complete the table of x- and y-values, and then plot and draw the graph of the equation.
Use the graph to read off the given values.

a) $y = \dfrac{3}{x}$

x	-2	-1	$-\frac{1}{2}$	$-\frac{1}{4}$	$\frac{1}{4}$	$\frac{1}{2}$	1	2
y								

Estimate the value(s) of y when $x = 2.3$.

b) $y = 3x + \dfrac{1}{x} + 2$

x	0.1	0.2	0.4	0.5	1	1.2	1.5	2
y								

Estimate:

   i) the value(s) of y when $x = 0.3$ and 1.6

   ii) the value(s) of x when $y = 5.8$ and 7.

c) $y = 2 - \dfrac{3}{x^2}$

x	1	2	3	4	5	6
y						

Estimate:

   i) the value(s) of y when $x = 1.5$

   ii) the value(s) of x when $y = 1.5$.

4 Copy and complete the table of x- and y-values, and then plot and draw the graph of the equation. Use the graph to read off the given values.

$y = 3^x - 1$

x	-2	$-1\frac{1}{2}$	-1	$-\frac{1}{2}$	0	$\frac{1}{2}$	1	$1\frac{1}{2}$	2
y									

Estimate:

a) the value(s) of y when $x = 0.8$ and 1.7

b) the value(s) of x when $y = 2.2$ and 6.

5 Imagine you have drawn the graph of the equation
$y = x^2 - 3x + 1$. Work out the equation of a straight line that
you must draw to help you solve each of these equations:

a) $x^2 - 5x + 1 = 0$ **b)** $x^2 - 2x - 4 = 0$

c) $x^2 + 3x - 1 = 0$ **d)** $2x^2 + 6x - 7 = 0$

6 a) Copy and complete the table of values for x and y, then plot
and draw the graph.

$y = 15 - 2x^2$

x	-3	-2	-1	0	1	2	3
y							

b) Use this graph to solve the equation
$15 - 2x^2 = 0$

7 a) Copy and complete the table of values for x and y, then plot
and draw the graph.

$y = x^2 - 5x + 5$

x	0	1	2	3	4	5
y						

b) Use this graph to solve the equation
$x^2 - 5x + 5 = 0$

8 a) Copy and complete the table of values for x and y, then plot
and draw the graph.

$y = x^2 - 3x$

x	-2	-1	0	1	2	3	4	5
y								

b) Use this graph to solve these equations.

    i) $x^2 - 3x = 4$

   ii) $x^2 - 3x = -1$

  iii) $x^2 - 3x - 7 = 0$

9 a) Copy and complete the table of values for x and y, then plot and draw the graph.

$y = 5 - 3x - 2x^2$

x	-3	-2	-1	0	1	2
y						

b) Use this graph to solve these equations.

i) $5 - 3x - 2x^2 = 0$

ii) $2 - 3x - 2x^2 = 0$

10 a) Copy and complete the table of values for x and y, then plot and draw the graph.

$y = 19 - 4x - 3x^2$

x	-4	-3	-2	-1	0	1	2	3	4
y									

b) Use this graph to solve these equations.

i) $19 - 4x - 3x^2 = 15 - 2x$

ii) $\frac{1}{5}(3x + 4) = \frac{5}{x}$

11 a) Copy and complete the table of values for x and y, then plot and draw the graph.

$y = x^3 - 3x^2 + 1$

x	-2	-1	0	1	2	3	4
y							

b) Use this graph to solve the equation

$x^3 - 3x^2 - 2x + 4 = 0$

An introduction to trigonometry

1 For each of these right-angled triangles (not drawn to scale), calculate a value for:

i) $\sin P$ **ii)** $\cos P$

iii) $\tan P$ **iv)** $\sin Q$

v) $\cos Q$ **vi)** $\tan Q$

Give your answers as fractions with rational denominators.

a)
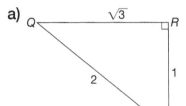

b)

c)
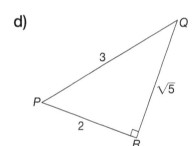

d)

2 In each case, $\triangle ABC$ is a right-angled triangle.
For each triangle:

i) construct the triangle accurately

ii) measure the length of the sides AC and BC

iii) calculate the sine, cosine and tangent for $\angle A$.

Give answers correct to 4 decimal places.

	AB	$\angle A$	$\angle B$
a)	10 cm	20°	90°
b)	8 cm	50°	90°
c)	6 cm	40°	90°
d)	3 cm	69°	90°

3 Use a calculator or tables to find these trigonometric ratios.
Give the answers correct to 4 decimal places.

a) $\tan 23°$ b) $\sin 45°$

c) $\cos 45°$ d) $\tan 25.5°$

e) $\cos 66.6°$ f) $\sin 12.1°$

g) $\tan 85.1°$ h) $\sin 70°$

i) $\cos 59.4°$ j) $\tan 0°$

k) $\cos 86.55°$ l) $\sin 2.36°$

m) $\tan 7.08°$ n) $\cos 15.43°$

o) $\sin 37.49°$

 4 Use a calculator and/or tables to evaluate each of these.
Give answers correct to 3 decimal places.

a) $\sin 36° + \cos 48°$ b) $6 \cos 52° + \sin 15°$

c) $15 \sin 44° \times 5 \cos 6°$ d) $10 \cos 42° - 3 \tan 11°$

e) $4 \tan 76° \times 3 \cos 75°$ f) $6 \sin 17.6° \div 3 \cos 4.6°$

g) $8 \cos 13.7° - 3 \tan 3.5°$ h) $7 \tan 72.4° + 3 \cos 13.8°$

i) $6 \tan 62.5° \div \cos 9.3°$

 5 If y is an angle between $0°$ and $90°$, find the value of y for each of
these trigonometric ratios, using a calculator or tables. Give the
answers correct to 1 decimal place.

a) $\tan y = 1.34$ b) $\sin y = 0.45$

c) $\cos y = 0.74$ d) $\tan y = 4.2$

e) $\cos y = 0.67$ f) $\sin y = 0.78$

g) $\tan y = 0$ h) $\sin y = 0.174$

i) $\cos y = 0.707$ j) $\tan y = 200$

k) $\cos y = 0.123$ l) $\sin y = 0.99$

m) $\tan y = 11.43$ n) $\cos y = 0.865$

o) $\sin y = 0.829$

 6 $\triangle ABC$ is a right-angled triangle as
shown in the diagram.

Use this diagram with a calculator or tables
to calculate each of the missing values in
the table at the top of page 45.

Give the side lengths correct to 2 decimal
places.

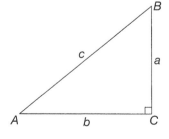

	$\angle A$	$\angle B$	$\angle C$	a	b	c
a)	52°	i)	90°	9.5 m	ii)	iii)
b)	50°	i)	90°	ii)	6.5 mm	iii)
c)	i)	64°	90°	12 cm	ii)	iii)
d)	i)	57.2°	90°	ii)	iii)	8 cm
e)	53.1°	i)	90°	4 cm	ii)	iii)
f)	31.9°	i)	90°	ii)	71.6 m	iii)
g)	i)	21°	90°	ii)	iii)	11 cm
h)	16°	i)	90°	ii)	15.1 mm	iii)
i)	i)	15°	90°	12.5 mm	ii)	iii)
j)	13°	i)	90°	ii)	iii)	1.4 m
k)	i)	27.7°	90°	ii)	18.1 cm	iii)
l)	i)	54°	90°	ii)	8.7 cm	iii)
m)	43.1°	i)	90°	ii)	iii)	24.2 mm

7 $\triangle ABC$ is a right-angled triangle as shown in the diagram.

Use this diagram with a calculator or tables to calculate each of the missing values in the table below.

Give the side lengths correct to 2 decimal places.

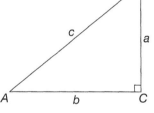

	$\angle A$	$\angle B$	$\angle C$	a	b	c
a)	36°	i)	90°	15.72 mm	ii)	iii)
b)	i)	53°	90°	12 cm	ii)	iii)
c)	i)	81°	90°	ii)	iii)	47.8 cm
d)	67°	i)	90°	ii)	15 m	iii)
e)	i)	61°	90°	17 mm	ii)	iii)
f)	i)	42.3°	90°	ii)	iii)	16 cm
g)	16.27°	i)	90°	ii)	35.46 m	iii)
h)	i)	58.7°	90°	ii)	iii)	172.6 cm
i)	i)	33.7°	90°	28.9 cm	ii)	iii)
j)	63.6°	i)	90°	32.14 mm	ii)	iii)
k)	31.23°	i)	90°	ii)	47.6 m	iii)
l)	28.7°	i)	90°	ii)	iii)	25.6 cm
m)	i)	47.9°	90°	ii)	32.65 cm	iii)

8 △*ABC* is a right-angled triangle as shown in the diagram.

Use this diagram with a calculator or tables to calculate each of the missing values in the table.

Give the answers correct to 3 significant figures.

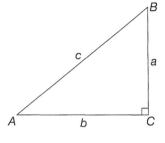

	∠*A*	∠*B*	∠*C*	*a*	*b*	*c*
a)	i)		90°	11 cm	27 cm	ii)
b)		i)	90°	ii)	14.5 m	17 m
c)	i)		90°	5.3 cm	ii)	8.6 cm
d)	i)		90°	ii)	17.6 cm	19.5 cm
e)		i)	90°	15.8 mm	ii)	21.2 mm
f)		i)	90°	ii)	4 m	6.7 m
g)	i)		90°	15.7 cm	ii)	19.3 cm
h)		i)	90°	3.5 cm	4.9 cm	ii)
i)		i)	90°	5.7 m	ii)	9.4 m

 9 Use the trigonometric ratios to calculate the unknown angles and/or sides in each figure.

a)

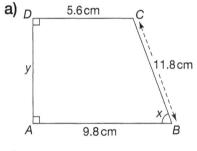

b)

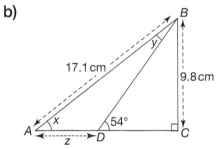

c)

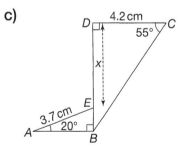

d)

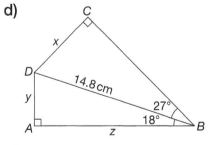

e)

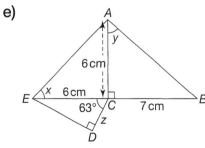

f)

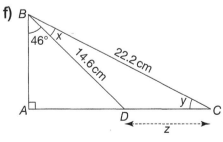

Give the answers to questions **10** to **22** correct to 2 decimal places for lengths and correct to 1 decimal place for angles.

10 A ladder that is 9 m long leans against the wall of a building at an angle of 55° above the ground.

How far from the building is the bottom of the ladder?

11 The driveway of a house slopes downhill from the road to the house.

The house is 2.6 m below the road and the driveway is 23 m long.

What is the angle of depression of the driveway?

12 A mobile phone aerial mast is held upright by four wires fixed to the ground.

Each of these wires is fixed 62 m away from the bottom of the mast and makes an angle of 34° with the ground.

Calculate the height of the mast.

13 An old oak tree is 10.4 m tall. On a sunny morning, the tree has a shadow that is 13.3 m long.

Calculate the angle of elevation of the sun.

14 A cliff is 31 m high. The angle of depression from the top of the cliff to a ship is 11°.

How far from the base of the cliff is the ship?

15 A man measures the angle of elevation from a point 53 m away from a house to the top of its chimney. He finds the angle to be 25.2°. He finds that the angle of elevation to the roof of the house is only 22.7°.

Calculate the height of the chimney.

16 Two buildings stand on level ground. One building is 114 m tall and the other is 79 m.

The angle of depression from the top of the taller building to the top of the shorter one is 37.1°.

How far apart are the buildings?

17 A helicopter leaves the base and flies 79 km on a bearing of 227°.

a) How far south of the base is the helicopter now?

b) How far west of the base is the helicopter now?

18 Car A is 24.8 km due west of a town. Car B is 14.3 km due north of the same town.

    a) What is the bearing of car A from car B?

    b) What is the bearing of car B from car A?

19 A fishing boat sails from the harbour on a bearing of 113°. The boat is now 67 km due east of the harbour.

How far has the boat sailed?

20 A plane leaves the airport and flies on a straight course for half and hour.

The plane is now 238 km to the south of the airport and 316 km to the west of the airport.

On what bearing did the plane fly?

21 A cruise ship sails 72 km on a bearing of 341°, and then 49 km on a bearing of 253°.

How far north is the cruise ship from its starting position?

22 A bird flies 2.6 km from its nest on a bearing of 071° and then flies 3.1 km on a bearing of 153°.

On what bearing must the bird fly to return directly to its nest?

An introduction to probability

1 Use one of the probability words 'impossible', 'unlikely', 'equal', 'likely' or 'certain' to describe each of these events.

  a) I choose a red card from a normal pack of cards.

  b) I choose a queen from a normal pack of cards.

  c) I roll a normal dice and get a score of 7.

  d) It will rain tomorrow in your town.

  e) I roll a normal dice and get a score divisible by 3.

  f) A coin is tossed ten times and it lands tails up six times.

  g) I choose a red bead from a bag that contains 12 blue beads and 15 green beads.

2 Here are the probabilities of four events.

Event W I choose a heart card from a normal pack of cards.
Event X I toss a normal coin and it lands heads up or tails up.
Event Y I roll a normal dice and get a score less than 6.
Event Z I choose a blue card from a normal pack of cards.

Copy this probability scale and label each of the events marked on the scale.

0                          $\frac{1}{2}$ 1

3 Here is a special dart board. A dart always lands on the board.

  a) List all the possible outcomes and write down n(*Total*).

  b) Event A is 'the dart lands on a vowel'.
   List all the outcomes of A, and write down n(A).

  c) List all the outcomes of A', the complement of A, and write down n(A').

4 A bag contains one silver button, two gold buttons and four black buttons.

A button is chosen at random from the bag.

a) List all the possible outcomes and write down n(*Total*).

b) Event *G* is 'a gold button is chosen'. List all the outcomes of *G*, and write down n(*G*).

5 A fair dice is thrown. Work out the probability that the number is:

a) not divisible by 3   **b)** bigger than 3   **c)** smaller than 1.

6 Here are some spinners. For each spinner work out
i) P(white) and ii) P(grey).

a) 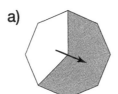 **b)** 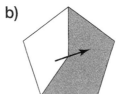 **c)** 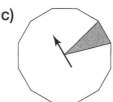 **d)**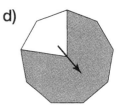

7 The pages of a thin exercise book are numbered from 1 to 42.
A page is chosen at random.
Write down the probability that the page number:

a) contains 2 digits

b) does not contain 2 digits

c) contains at least one digit of the number '2'

d) is a perfect square.

8 A box contains 4 yellow marbles, 3 green marbles and 3 purple marbles.

A marble is removed from the box at random.
What is the probability that the marble is:

a) yellow   **b)** green or purple

c) yellow, green or purple   **d)** red?

9 A card is chosen at random from a normal pack of playing cards without jokers and with an ace counting as a one.
Find the probability that this card is:

a) a red picture card   **b)** not a red picture card

c) a black card lower than 8   **d)** a red six or a black three.

(Write these probabilities as percentages correct to 2 d.p.)

10 A spinner can land on red, green, blue or yellow. The spinner is spun once.

The probability that it lands on red or blue is 50%.

The probability that it lands on red or green is 55%.

The probability that it lands on blue is 20%.

What is the probability that it lands on:

a) red   **b)** green   **c)** yellow?

11 A bag contains 36 blue pens and some red pens. One pen is chosen at random.

The probability of choosing a red pen is $\frac{4}{13}$.
Calculate the number of red pens in the bag.

12 Some green counters and some yellow counters are labelled either X or Y as shown:

	X	Y
Yellow	14	16
Green	34	16

One of the counters is chosen at random.

What is the probability that this counter is:

a) an X   **b)** green   **c)** a green X?

A green counter is chosen at random.

d) What is the probability that this counter is an X?

A counter labelled X is chosen at random.

e) What is the probability that this counter is green?

13 Here is a fair spinner with six equal sections numbered 1, 1, 2, 2, 3 and 4. Nicco spins twice and adds his scores together.

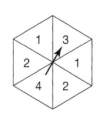

Draw a possibility diagram to work out the probability of:

a) scoring 5

b) scoring less than 5

c) scoring an odd number.

14 Bag A contains counters numbered 2, 4 and 6. Bag B contains counters numbered 3, 5 and 7. A counter is chosen at random from Bag A and then a counter is chosen at random from Bag B. The value of the counter from Bag A is divided by the value of the counter from Bag B.

Draw a possibility diagram to show all the possible outcomes.

Find the probability that:

a) the quotient is less than $\frac{1}{2}$

b) the quotient is bigger than 1

c) the quotient is bigger than $\frac{1}{2}$ but smaller than 1.

15 Two normal fair dice are thrown together.

Draw a possibility diagram to show all the possible outcomes for the two dice.

Find the probability that:

a) the sum of the two numbers is 12

b) the sum of the two numbers is 5

c) the score on one dice is exactly double the score on the other.

16 A fair dice is rolled and a fair coin is tossed at the same time.

Draw a possibility diagram to show all the possible outcomes.

Find the probability of getting:

a) an odd number   **b)** a head and a 6

c) a tail and an odd number   **d)** a head and a number bigger than 2.

17 Two spinners each have four equal sections. The sections on spinner P are numbered 1, 2, 2 and 3. The sections on spinner Q are numbered 1, 2, 5 and 6. Each spinner is spun once.

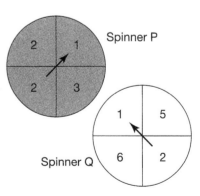

Draw a possibility diagram to show all the possible outcomes.

Find the probability that:

a) one score is a multiple of the other

b) the total score is a prime number

c) the total score is less than 7

d) the total score is a perfect square.

18 Ben throws a special six-sided dice (not a normal one). The probability of throwing a six is $\frac{1}{3}$. The probabilities of throwing a two, three, four or five are each $\frac{1}{6}$.

a) What is the probability of throwing a one? What does this tell you about the numbers on Ben's special dice?

Ben now throws his special dice twice and adds the two scores.

b) Draw a possibility diagram for all the possible outcomes.

c) What is the probability that the sum of the scores is six?

d) What is the probability that the sum of the scores is twelve?

19 100 teenagers are waiting to audition for a show. 64 of them are boys. The order of the auditions is chosen at random.

Find the probability that:

a) the first teenager to audition is a girl

b) the first two teenagers to audition are both girls

c) the first two teenagers to audition are not both girls.

Use a tree diagram to help you work out the probabilities.

20 Jamelia has some coins in her purse. She has four 50¢ coins and two 20¢ coins.

She takes out two coins at random from the purse.

Draw a tree diagram to show all the possible outcomes.

Find the probability that:

a) the total value of the two coins is 40¢

b) the total value of the two coins is 70¢

c) If Jamelia takes out a third coin at random, find the probability that the total value of the three coins is 90¢.

21 Simon finds it hard to wake up early in the morning. He sometimes oversleeps and is late for school. The probability that he is late on any day is 0.3.

Draw a tree diagram to work out all the possible outcomes on two consecutive days.

Find the probability that:

a) Simon is late for school on both days

b) Simon is late only once.

22 Two teachers, five girls and three boys want to join the school sports committee. There are only two places vacant on the committee, and the successful candidates are chosen at random.

Draw a tree diagram to work out all the possible outcomes.

Find the probability that:

a) both candidates are boys

b) neither candidate is a boy

c) one candidate is a boy and the other is a girl

d) at least one of the candidates is a teacher.

23 Spinner A has sections coloured yellow, green and purple as shown. Spinner B has sections coloured yellow and green as shown. Each spinner is spun once.

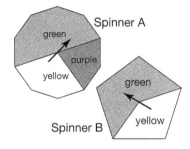

Draw a tree diagram to show all the possible outcomes.

Find the probability that:

a) spinner A lands on yellow and spinner B lands on green

b) the two colours are green and purple

c) the two colours are the same.

24 The probability that is rains on any day during August in Sita's town is 0.4. If it rains, the probability that she takes the bus to school instead of walking is 0.7. If it does not rain, the probability of Sita walking to school is 0.9.

Draw a tree diagram to work out all the possible outcomes.

Find the probability that:

a) Sita takes the bus to school when it is not raining

b) Sita walks to school on a day in August.

25 The probability that Rangers Rugby Club will win a match is $\frac{3}{4}$. Find the probability that:

a) the club will not win the first match of the season

b) the club will win one of the first two matches of the season.

The probability that the team will draw any match is $\frac{1}{6}$.

c) Find the probability that the club will lose the third match of the season. (Hint: {not win} = {lose + draw})

Unit 12 Statistical averages

1 Calculate the mean of each of these sets of data, correct to 1 d.p.

   a) 5, 5, 5, 7, 7, 8, 8, 8, 8   **b)** 46, 67, 68, 41, 56, 65, 77

   c) 8, 7, 5, 9, 3, 6, 11   **d)** 11.9, 12.6, 12.0, 13.6, 12.9, 11.4

   e) 316, 317, 318, 321, 323   **f)** 110.5, 119.6, 114.2, 122.4, 132.3

2 The mean age of three children is 12 years. If two more children join the group, the mean age is now 10 years.

   Calculate the mean age of the two new children.

3 The mean of five numbers is 46. Two of the numbers are 29 and 96. Find the other three numbers if they are equal.

4 The mean mass of 20 apples and 14 pears is 161 g. The mean mass of the pears on their own is 151 g.

   Calculate the mean mass of the apples on their own.

5 The mean of eight numbers is 3. When twelve more numbers are included, the mean becomes 9.

   Calculate the mean of these twelve numbers on their own.

6 Karl thinks that the boys in his class wear bigger shoes on average than the girls. He asks each child in his class what their shoe size is. His results are shown in the table.

   Calculate the mean shoe size for the boys and for the girls to decide if Karl is correct.

Shoe size	37	$37\frac{1}{2}$	38	$38\frac{1}{2}$	39	40	41	42	43	44	45
No. of boys	0	2	0	2	3	0	2	0	3	1	1
No. of girls	1	0	5	4	4	2	1	0	1	0	0

7 The amount of vitamin C was measured in some grapefruit. The results are shown in the table.

   Calculate the mean amount of vitamin C in the grapefruit correct to 3 s.f.

Vitamin C (mg)	26	27	28	29	30	31	32	33	34
No. of grapefruit	9	8	15	21	23	13	9	1	3

8 Sam asks some pupils at his school how many bags of crisps they eat in a week.

The results of his survey are shown in the histogram.

Calculate the average number of bags of crisps that these pupils eat in a week, correct to 1 d.p.

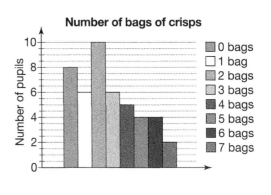

9 Some pupils took a history quiz. Their scores are shown in the histogram.

Calculate the mean score for these pupils in the quiz, correct to 2 s.f.

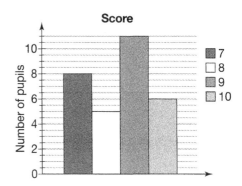

10 A factory makes cotton buds.
Each pack should contain 50 cotton buds.
The contents of some sample packs were counted.
The results are shown in the stem-and-leaf diagram.

Number of cotton buds

4 | 8 8 8 8 9 9 9 9 9 9 9 9 9
5 | 0 0 0 0 0 0 0 0 0 0 0 0 0 0 1 1 1 1 1 1 1 1 2 2 2 2 2 2 2 2 2 2 3 3 3

Calculate the mean number of cotton buds in a pack.

11 For a biology project, some pupils measured the heights of some plants.

Their results are shown in the stem-and-leaf diagram.

Calculate the mean height of the plants correct to 1 d.p.

Height in cm

2 | 5 8
3 | 0 1 1 3 5 6 7 7 8 8 9
4 | 0 0 0 1 1 3 3 3 4 4 5 6 6 7 7 8 9 9
5 | 0 1 2 2 3 4 6 8 9

12 Calculate the median of each of these sets of numbers.

a) 7, 10, 9, 12, 11, 11, 8, 7, 9

b) 56, 54, 58, 60, 67, 70, 49, 53

c) 7.3, 6.7, 4.8, 3.3, 6.7, 5.5, 4.9, 6.6, 4.3

d) $5.86, $10.34, $16.92, $12.98, $14.73, $9.85

13 The median of a set of eight numbers is 39. Seven of the numbers are 35, 51, 43, 34, 38, 35 and 47.

Find the eighth number.

14 The table shows the number of times the fire brigade was called out each day over some weeks.

Calculate the median number of call-outs in a day during this period.

No. of call-outs	0	1	2	3	4	5	6	7
No. of days	13	9	8	14	10	3	1	2

15 Some pupils counted the number of cars that drove through a road junction every minute over a period.

The histogram shows their results.

Calculate the median number of cars per minute that drove through this junction.

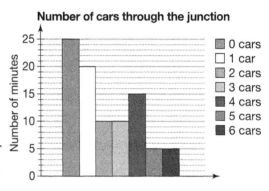

16 A Year 9 class write tests in English and mathematics. The table shows their scores.

Calculate the median scores for English and for mathematics in these tests.

Scores	1	2	3	4	5	6	7	8	9	10
No. of pupils (English)	1	6	13	3	8	3	4	1	1	0
No. of pupils (maths)	3	2	6	4	5	12	3	1	3	1

17 Year 9 pupils in Flowerville School measured the heights of some Year 7 pupils.

Their data is shown in the stem-and-leaf diagram.

Calculate the median height of the Year 7 pupils measured.

Height in cm

```
14 | 4 0 7
13 | 6 6 9 2 8 5 7 5 3 0 1 3
12 | 2 6 0 5 9 7 4 2 9 8 7 8
11 | 6 8 9
```

18 Find the mean, median and mode of each of these sets of data.

a) 1, 1, 2, 3, 5, 8, 1, 3, 2, 1

b) 5, 3, 7, 3, 3, 9, 0

c) 8, 11, 14, 13, 14, 9, 15

d) 12, 11, 13, 11, 15, 16

e) 0, 5, 5, 1, 4, 1, 6, 7, 3, 0, 1

f) 53, 5, 23, 58, 6, 20

g) 88, 93, 85, 98, 102, 98, 93, 104, 102, 98

h) 555, 607, 109, 990, 709, 384

i) 14.3, 13.5, 10.5, 12.6, 15.3, 16.4, 12.6, 16.0

j) 86 kg, 95 kg, 89 kg, 93 kg, 84 kg, 78 kg, 91 kg, 79 kg

19 For each of these frequency distributions, find
i) the mean, ii) the median and iii) the mode.

a)

Value	150	155	160	165	170	175
Frequency	2	4	3	5	2	1

b)

Value	42	47	52	57	62	67	72	77
Frequency	5	23	32	23	11	3	2	1

20 A normal fair dice is thrown a number of times and the results are shown in this table.

Score	1	2	3	4	5	6
No. of throws	12	9	11	6	4	7

a) Find the median and the mode of the scores.

b) The dice is thrown one more time and the mean of all the throws is exactly 3.

What is the number on the last throw?

21 Pupils in Flowerville School were asked how many people live in their house. Their answers are shown in the frequency distribution table.

No. of people	2	3	4	5	6	7	8	9
Frequency	58	206	284	54	45	52	76	25

a) Calculate the mean number of people living in a house.

b) Find the sum of the mode and median values for this data.

22 The numbers 3, 7, 13, 14, 16, 19, 20 and x are listed in ascending order.

The mean of these numbers is equal to the median. Find the number x.

23 Mr Abraham sows 6 seeds in each plant pot.

When the seeds start to germinate, not all of them grow into healthy plants. Mr Abraham counts the number of healthy plants in each pot.

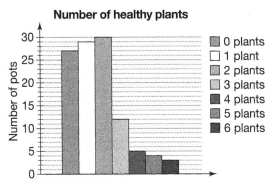

His results are in the histogram. Find:

a) the mean number of healthy plants

b) the median number of healthy plants

c) the modal number of healthy plants.

24 A shopkeeper records how many litres of milk she sells each week for a period.

The results are shown in the stem-and-leaf diagram.

Litres of milk sold

```
3|4 6 0 1
2|3 7 0 5 1 8 0 2 4
1|2 7 4 8 9 5 1 0 3 1 6 6 4 3 9 3 6 8 1 2 4 7 0 2 8
0|4 9
```

Find:

a) the mean number of litres sold per week

b) the median number of litres sold per week

c) the modal number of litres sold per week.

25 Some people are asked how much money they earn each month. The results are shown in the histogram.

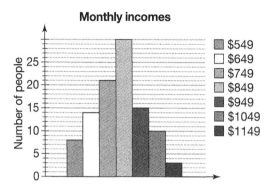

Monthly incomes

Legend:
- $549
- $649
- $749
- $849
- $949
- $1049
- $1149

Find:

a) the mean monthly income

b) the median monthly income

c) the modal monthly income.

26 A normal fair dice is thrown 20 times and the results are shown in this table.

Score	1	2	3	4	5	6
No. of throws	5	4	a	b	3	2

The modal score is 4. Find the value of a and b.

27 Pupils at Flowerville School wrote a science exam.

The time taken by each pupil to complete the exam was recorded and these are shown in the stem-and-leaf diagram.

Minutes

```
5|1 2 7 7 2 3 4 8
4|9 2 1 4 1 2 8 1 1 5 6 2 3 4 7 7 2 7 3 6 2 1 4 2 2
3|5 6 6 5 1 3 9 8 5 5 9 1 1 2 5 1 1 9 1 3 8
2|1 5 2 8 7 5
```

Find:

a) the mean number of minutes

b) the median number of minutes

c) the modal number of minutes.